Woodpecker and Raven

A Bird Book for Kids™

By Novare Lawrence

Nada Bindu Publishing Co.

The contents of this book previously appeared in the digital-only e-book editions *Woodpecker: A Bird Book for Kids* and *Raven: A Bird Book for Kids*.

First Print Edition – June 2018

ISBN-10: 1-63307-017-4
ISBN-13: 978-1-63307-017-2

Published by:
Nada Bindu Publishing Co.
Cheyenne, WY 82001
Website: www.nadabindupublishing.com
Email: inquiries@nadabindupublishing.com

To My Readers

Woodpecker and Raven *is my fifth paperback book, a format which highlights the truly beautiful pictures of these birds and their lives. These two birds first appeared as their own e-books which have been combined without change here into this printed volume. The number of e-books in the **A Bird Book for Kids** series now numbers twelve and counting.*

I invite your comments and reviews on the website where you bought this book. This will help me to continue to create books of the highest quality and enjoyment for all my readers. Along with your review, please let me know what other birds you would like to see as I continue to expand this series, as e-books first and as paperback editions second that combine two e-books into a single printed book.

Many Thanks,
Novare Lawrence

CONTENTS

Woodpecker

Woodpeckers are small birds that pack a big punch. With their strong pointed beaks and extra support in their skulls, woodpeckers are able to drill into tree bark and tree trunks to find and extract insects. They can also excavate deep holes that they use as their nests.

Outside of a nest

Most birds have specialized beaks or bills which improve their ability to feed themselves. Some birds, like hawks, eagles, and owls, have hooked beaks suited for removing the fur and skin of the prey that they eat. Others, like parrots and finches, have beaks suited for cracking nuts and seeds. Hummingbirds have extra-long bills that allow them to reach into flowers to drink nectar using their equally long tongues.

Four different beaks for different food: finch, eagle, hummingbird and woodpecker

Woodpeckers hammer their special bills into a tree trunk or branch very fast, at speeds up to 15 miles per hour. They hit the wood with the force of 1000 gravities. Humans can only withstand a maximum of about 10 gravities and for that they wear special suits that help to keep blood in the brain so they don't pass out. Woodpeckers generate this incredibly strong force each time they hammer with their beak to create a hole in a tree trunk.

A Gila Woodpecker digging out an insect

Woodpeckers have some unique anatomy in order to survive their pecking activity. Their bills have a chisel-like tip and a special inner bone layer that helps to absorb the shock of the impact against the wood. Their brains also have very little fluid between their brain and skull unlike the way that humans do. This means that their skull acts like a bicycle safety helmet would for us. It holds their brain tightly, absorbing the shock while keeping the brain from moving inside the skull and being injured.

A Great-spotted Woodpecker: strong beak and compact skull

Once a woodpecker's bill penetrates deeply enough into the wood to find a burrowing insect, the woodpecker extends their long tongue into the hole. Their tongue is special too. Beside its extra length, it is barbed so that it can grab the insects or grubs that it finds and pull them out for a tasty meal. The woodpecker is perfectly adapted to find and eat food that other birds cannot reach.

A Great-spotted Woodpecker with a successful catch

Woodpeckers are the largest group of a big family of birds called the Picidae. This family includes almost 200 different species of related birds like woodpeckers, piculets, wrynecks, and sapsuckers. One common thing about most Picidae birds is that they have feet where the inner two toes point forward and the outer two toes face backward. This makes it easier for them to grab onto a tree trunk and stand vertically. There are, however, a few species of three-toed woodpeckers where only one toe points backwards.

Two toes forward and two toes backward: a Black-headed Woodpecker

There are over a 150 species of what are called true woodpeckers. The smallest woodpecker in the United States is the downy woodpecker. The adult downy is from 5-1/2 to 7 inches (14-18 cm) in length and from 0.75 to just over 1 oz. (20 – 33 g) in weight. Their wingspan averages around 2-1/2 feet (76 cm).

A male Downy Woodpecker

The downy woodpecker lives mostly in deciduous forests, which are trees that lose their leaves in the fall and winter months. They may migrate to avoid harsh winters but generally live in the same area year-round. They search for insects in the tree bark and trunk but also eat seeds and berries, especially in the winter time. People can attract downy woodpeckers, among other birds, by putting a mesh bird feeder with suet in it. Suet cakes usually consist of oats, bird seed, cornmeal, nuts and raisins held together with lard.

A female Downy Woodpecker at a suet feeder

The largest woodpecker is the 12 to 20 oz. (340 to 567 g) great slaty woodpecker which is found in South Asia, from Nepal, to India and the Philippines. It can reach almost two feet (61 cm) in length. The bill itself is 2.4 to 2.6 inches (6 to 6.5 cm) long. Groups of these woodpeckers may band together to find food. Their preferred meal is to dig into insect nests which contain a lot of bugs. This includes nests of ants, termites, beetles and non-stinging bees. Sometimes they will also feed on small fruits.

A Pileated Woodpecker – the second largest woodpecker in North America

Beyond the true woodpeckers are the closely related Piculets which are the smallest in the Picidae family. The rufous piculet, for example, is just 3 to 4 inches (7.6 – 1o cm) and has almost no tail feathers. They weigh only about 1/3 of an ounce (9.2 g) so they are not able to drill into tree trunks the way their woodpecker cousins can. Instead, they use their pointed beaks to probe into existing holes on tree trunks and branches or into decaying wood where insects are living and scoop them out with their long tongues.

A Rufous Piculet

There are about thirty species of piculets. Most live in the tropical areas of South America but there are a couple species found in Asia and one in Africa. Because piculets have almost no tail feathers to help them balance, they generally prefer to sit on top of a branch than to hold on to a tree trunk and stand upright like true woodpeckers. When gathering food, however, they are quite active and can be seen circling up a tree trunk as they search for insect holes and a meal.

A White-browed Piculet

There are two species of wrynecks, close in size to the smaller true woodpeckers. Wrynecks do not have the stiff tail feathers that woodpeckers use to balance themselves on a tree trunk while they hammer the wood with their beak. This means that they cannot generate the same drilling force. Wrynecks will use their pointed beaks mostly on decaying wood that is soft or digging in the soil to find insects and grubs to eat. Wrynecks are found only in Asia and Africa.

A Wryneck

Sapsuckers are another Picidae species related to woodpeckers. As their name suggests however, insects are not the main reason for drilling. They drill holes in trees for the sap which provides two food items for the birds. Insects are attracted to the sweet sap that comes out these holes and get stuck. Sapsuckers then will eat both the insects and the sap. During breeding season, and when the young have hatched, the sapsucker will take these insects to feed their young. Sapsuckers may also eat seeds and berries.

A Sapsucker checking holes for sap and insects

There are four species of sapsuckers. They live in North American forests and woodlands, from Alaska, across Canada and the United States and down to Mexico. Because sapsuckers drill lots of holes in a tree in order to get sap, they can actually weaken a tree's health. For this reason, some people consider sapsuckers to be pests while woodpeckers may be more beneficial since they actually eat insects that have burrowed into trees.

A Williamson Sapsucker on a pine tree

Flickers are another group of birds associated with woodpeckers. There are nine types of flickers and they generally prefer foraging for food on the ground. They use their strong beaks to dig into the earth or into insect nests like ants and termites. Flickers are found in North, Central and South America and among the Caribbean islands, including Cuba. The northern flickers usually migrate to the south for winter while the southern flickers tend to live in the same area all year round. While woodpeckers almost always nest in tree trunks, some flickers may use old ground nests abandoned by other birds or animals.

A Flicker in flight

Most woodpecker species mate for life. During the breeding season, male woodpeckers are usually responsible for creating the nest by boring a hole into a tree trunk. The opening is just large enough to squeeze through but the hollow inside the trunk must be big enough for up to 7 eggs and the parent that incubates them. The female sits on the eggs during the day while the male does it at night so the eggs are never left alone. Both parents can be very aggressive in defending their nest.

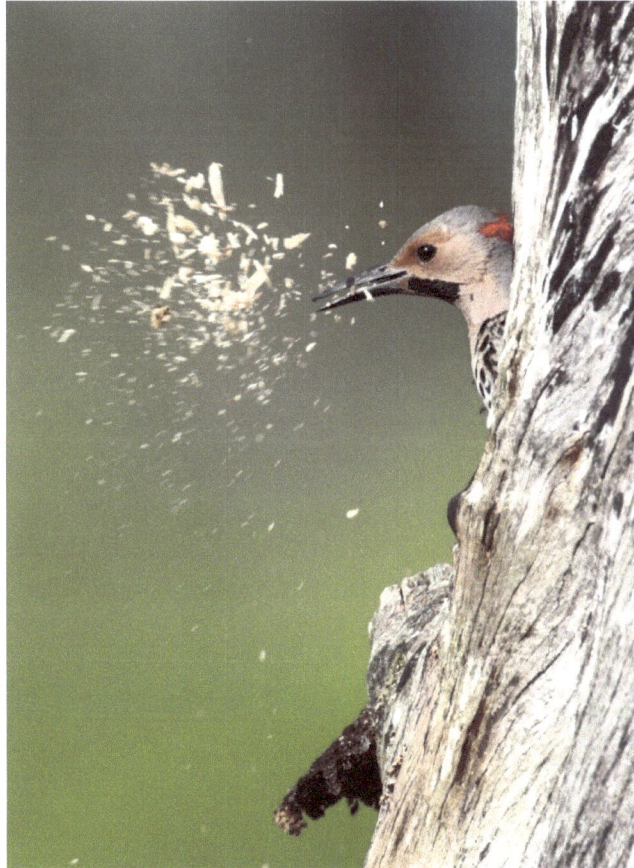

A woodpecker hollowing out a nest

It takes almost a month to hammer out a nest inside of a tree trunk so woodpeckers can be quite territorial. They will aggressively defend their nest as other woodpeckers and birds will try to use it for their own eggs. Some woodpeckers may re-use the same nest each year but it is more common that they either create a new one or re-use one created by another bird. Woodpeckers and piculets are able to create their own nests but wrynecks do not. Wrynecks must use one created by another woodpecker.

A female Spot-Breasted Woodpecker in the nest

The number of eggs that the female woodpecker lays is different depending on her species. Most will lay between 2 to 7 eggs. Because woodpeckers nest inside trees, their eggs are white. The white eggs are easier for the parents to see in the dimly lit hole which is their nest. Other types of birds build nests out of twigs or grasses on tree branches or on the ground and they have colored or spectacled eggs. The coloring and dotted patterns on the egg are meant to help hide the eggs from being seen by predators but woodpecker's eggs set deep in a tree trunk don't need these colors or patterns for safety.

Two hungry woodpecker chicks

Woodpecker chicks hatch within two weeks. Both parents help with feeding the chicks. It takes from 18 to 30 days for the chick to fledge and fly out of the nest. The young chick may spend some time with the parents after fledging but will then go off on its own. Woodpeckers mostly are solitary but young birds may form informal groups with other young woodpeckers of the same species until they mature and begin looking for a mate.

A young Great-Spotted Woodpecker almost ready to fledge

Woodpeckers, which are non-passerine birds, have a more limited series of calls than passerine birds which are also called songbirds. Woodpeckers use courtship calls and alarm calls, including those sounded during territorial disputes. Mated pairs can recognize each other's calls so that they can find each other within the dense forest. But woodpeckers have a special communication method that songbirds don't – woodpeckers drum.

A female Red-Bellied Woodpecker landing

Woodpeckers searching for food hammer in a specific pattern: usually about four hits with their beak against the tree trunk, then a pause, then four more hits, and so on. When they drum, they are not trying to drill a hole to find food. Instead, they look for something that resonates, like a hollow tree or even man-made pipes, and begin a drumming pattern unique to their particular species. This drumming song is used to identify a territory but may also play a role in courtship. Male woodpeckers are known to drum more than female woodpeckers.

A male Black Woodpecker preparing to drum

While there are a small number of woodpeckers such as the Gila woodpecker that is found in the desert and lives in taller cactus plants, woodpeckers depend on trees and forests. Their ability to find food and create safe nests in which to raise their young all depend on having trees. Humans also have a need for trees and we cut them down to use their wood and resins for many purposes including cooking food, building shelters and furniture, and making many paper products.

A Gila Woodpecker on a palm tree trunk

Humans try to support many kinds of birds and animals by both setting aside and protecting some forests from any logging and by replanting programs in planned logging areas. This allows birds like the woodpeckers, some species of which are already classified as threatened, to continue to have areas in which they can live and raise their young. This means that now, and in the future, we should still be able to hike through a forest and hear those distinctive hammering patterns and perhaps even see a woodpecker as he circles a tree trunk to find a bit of food.

A European Green Woodpecker digging for insects

Raven

Ravens are small to medium sized birds with big personalities. Covered in black glossy feathers, sporting a strong, thick beak and known for their distinctively loud and frequently used "caw" call, ravens are smart birds that stand up for themselves against all kinds of challenges.

Two ravens ready for anything

The common raven, also called the northern raven, is found throughout the northern hemisphere, including North America, Europe and Asia. As its name suggests, it is the most commonly found raven.

A raven being curious

There are actually nine types of ravens found around the world. The common raven, thick-billed raven, white-necked and brown-necked raven are the larger ravens, weighing around three pounds (1.4 kg) and measuring about 25 inches (63.5 cm) long. These birds are among the largest passerine birds, which are birds with three toes on the front of their foot, and one toe in back. This arrangement of toes allows them to strongly grip branches so they live and nest primarily in trees where it is safer, rather than on the ground.

A white necked raven

The smaller ravens include the Chihuahuan raven, forest raven, little raven, fan-tailed raven and Australian raven, which are all less than 20 inches (51 cm) long and only about 1.4 pounds (635 g). Ravens, particularly younger or smaller ravens, are often mistaken for crows, another black bird with a similar sounding call. Crows are smaller and have a thinner beak than ravens. It is not always easy to tell the difference between crows and ravens at a distance.

An Australian raven on the beach eating a fish

Ravens, along with crows, magpies, rooks, jays and nutcrackers are part of one very large family of birds called the corvids. One special aspect of this large family of birds is that they are considered to be the most intelligent of birds and among the smartest of all animals. Ravens and crows often learn how to use tools in order to get food. For example, they may use sticks to get bugs or seeds that they can't reach with just their beaks.

Looking for something in the trees

Common ravens can live to be over 20 years old in the wild which is the longest of all ravens and most other passerine birds. The oldest raven, bred to be one of the six ravens which live on the Tower of London grounds in England, is known to have lived to the age of 44.

A raven searching for a bite to eat

In addition to being larger and longer lived than most passerine birds, common ravens can live comfortably in a greater variety of climates. Ravens can live in the cold Arctic and at high altitudes, such as in Tibet at an elevation of 20,800 feet (6350 meters) above sea level. They live throughout the varied climates in North America, Europe and Asia and in the deserts of North Africa. Ravens only migrate to escape the very coldest areas during winter, otherwise they live within the same general area all year round.

A raven in the snow

Two reasons that common ravens can so successfully live in such different climates is because of their diet and their very intelligent and opportunistic personalities. Ravens can hunt small reptiles and mammals, take other bird's eggs, and they also scavenge like vultures, eating meat from dead animals. They will also eat berries and other fruit, as well as grains and seeds. Their opportunistic nature is apparent as ravens also enjoy finding food waste from humans, much like pigeons and sea gulls.

A white necked raven sharing a meal with a vulture

Ravens display a lot of their intelligence in matters pertaining to food. Ravens often store food for later, hiding it from other ravens and birds. They also try to watch other ravens to discover these hidden food storage places. Ravens will follow other animals as well to take advantage of any food they may find. For example, ravens may follow a wolf or fox in the wintertime to pick up scraps from the leftovers that these predators may leave.

A raven with a mouthful

Male and female ravens look the same, unlike many other bird species such as peacocks and ostriches where males and females vary in both size and color. Juvenile ravens usually stay together in small groups but once a male and female choose each other, they will establish a breeding area. Ravens mate for life and are very protective of their chosen breeding area.

A pair of white necked ravens

Thick-billed ravens are found in eastern Africa. They are the largest of ravens and marked with a white patch on the back of their head and neck. Thick-billed ravens nest in trees and on cliffs. In their nest built from sticks, the female lays three to five eggs. They may reuse the nesting spot every year.

A thick-billed raven

Common ravens build more complex nests. Starting with sticks and twigs, they add mud, roots and even tree bark before a final soft layer, usually animal fur, is laid down. Trees and cliffs are the most common place for the nest but they may use phone or power poles and even abandoned buildings if necessary. The female lays three to seven eggs which hatch after about 18 days.

Raven chicks calling for food

The Chihuahuan raven, found in the U.S. southwest and Midwest, is one of the smaller ravens. These ravens may build their smaller nest in shrubs as well as in trees, utility poles and abandoned buildings. They lay five to seven eggs.

A hooded crow, closely related to ravens, is feeding her young hatchlings

Raven chicks hatch after eighteen to twenty-one days. The female does most of the incubating of the eggs but both parents will bring food for the chicks. It takes five to six weeks for the chicks to grow all of their feathers and to be able to leave the nest. Fledged young will often stay with their parents for up to six months before joining a group of juvenile ravens.

Young chicks almost ready to fledge

Ravens are very talkative with each other and have many different calls. Types of calls include alarms for predators or intruders, chase calls, calls when flying and location calls used in finding to a lost mate.

Ravens flying together

One of a raven's special abilities, like a parrot, is its ability to mimic sounds, including both environmental sounds and human voices. It has been observed in the wild that they can mimic foxes or wolves and lead them to a carcass the raven has found. This is important because ravens don't have the sharp, hooked beak of a raptor like a falcon or eagle so they can't cut through animal hair and skin. A fox or wolf can so the raven waits until they are done and happily swoops in to finish the exposed remains.

A raven interrupts a vulture's meal

Ravens can mimic human voices just as well as parrots or mynah birds do. In the wild, they rarely have enough exposure to people but in captivity, they can and do pick up human words and even household sounds. Of all the "talking" birds, parrots and parakeets seem to be able to memorize the most words and sounds.

Two other "talking birds" – the mynah and grey parrot

Ravens may also communicate with other ravens without using calls. They can snap their bills shut to produce a clapping or clicking sound. They can also produce what is called a "wing whistle." This flapping activity when they take flight produces a "whoosh" sound that usually communicates an alarm warning to other ravens.

Two ravens playing in the snow

Ravens have also been observed pointing to objects they want a fellow raven to notice. Or they may pick up an object, such as a twig, and wave it to get the other bird's attention.

Up in the tree together

One thing that ravens share with humans is that they like to play. They are curious and playful, just like any kid. Ravens will play in the snow by rolling or sliding down a snow-covered slope. They may pick up pine cones or even golf balls and then play with them by rolling them around and chasing them.

A raven playing in the water

Ravens play while flying, especially juvenile ravens. They show off their flying abilities by flying in tight loops or flying upside down. They may even lock talons with another raven and fly together spinning in a circle. Eagles and hawks also do this.

A white necked raven flying with his talons down

Ravens may tease other animals too just because they want to play. They might drop things like sticks or rocks repeatedly on a dog or pick up a stick and poke at it. It's sort of like playing tag but it's mostly to get the animal to chase them because they enjoy it. Ravens are quite quick and skillful flyers and seem to have little worry of being caught.

A raven hopping on the ground

Ravens are very good at solving problems, especially where food is involved. In different parts of the world, ravens have learned to drop nuts from above onto a road where cars will run over the nuts. The ravens then just have to swoop down to eat the soft meat inside. They've learned to pull up fishing lines to eat the fish or the bait attached to it. And they can use tools to get bugs or other food out of tight spaces in order to eat it.

Looking for leftovers in the park

With all the intelligence and curiosity that ravens show, it's no wonder that throughout history, ravens have long been part of myths and legends. They were often seen as messengers from gods, sometimes bringing good omens, sometimes bringing evil ones. Tibet, ancient Greece, the Celts, Vikings, ancient Chinese, and Native Americans all associated ravens with the gods.

A raven close up

Much like black cats, ravens, with their black feathers, piercing "caw" call, and their direct, intelligent eyes, became associated in European legends with graveyards and evil things. This is how ravens became one of the traditional symbols of Halloween.

Does he make you think of Halloween?

England has perhaps the most famous and still relevant legend concerning ravens. On the grounds of the Tower of London today, there are six ravens that live there full time. It is said that there has been a captive stock of ravens since around 1680 when a royal astronomer reportedly told the king that if the ravens were removed, the king and the entire kingdom would fall and be conquered. Since then, and even today, if you go to London, you will find six ravens on the grounds being cared for by the government. They, of course, have backup ravens in reserve. Just in case.

A special raven in London

ABOUT THE AUTHOR

Novare Lawrence loves researching and writing books about Nature. She shares the knowledge and beauty of our natural world with kids young and old hoping that we will all do our part to help preserve our planet and all the wonderful species upon it.

You may learn more about her books and the *A Bird Book for Kids*™ series at her website:

ABirdBookforKids.com

And at NadaBinduPublishing.com

A Bird Book for Kids ™ Books   by Novare Lawrence

Digital:

Paperback:

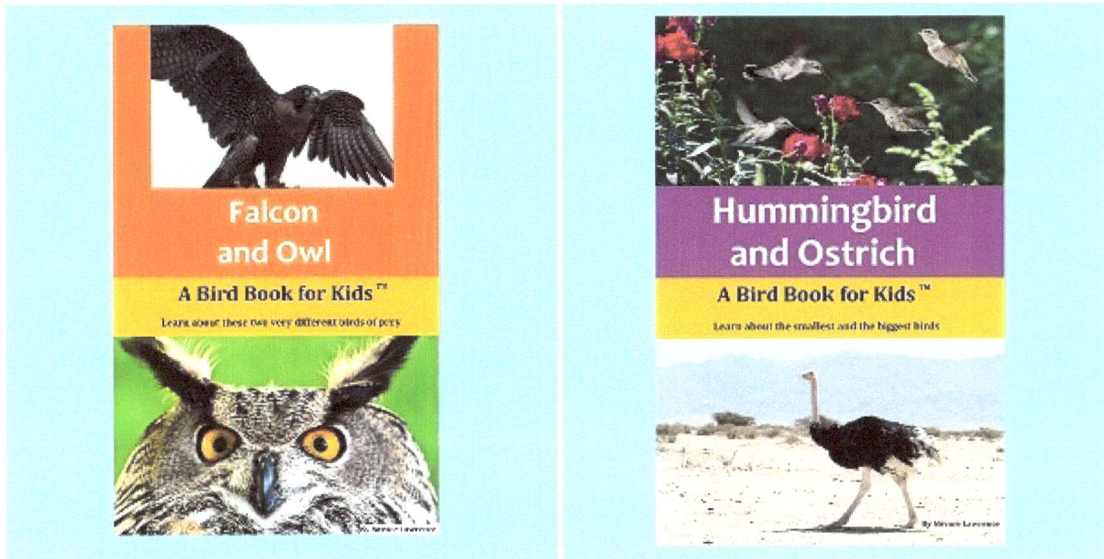

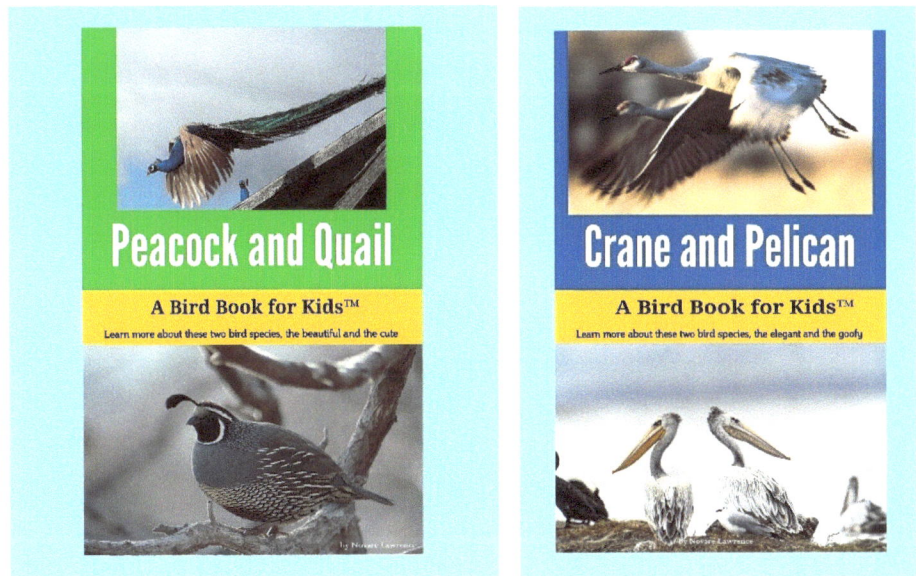

www.ingramcontent.com/pod-product-compliance
Lightning Source LLC
Chambersburg PA
CBHW060817270326

41930CB00002B/68